Clever CEREAL Crafts

Publications International, Ltd.

Amy Belonio has an M.A. in art education and has taught art for five years; she is also a freelance designer and a very busy mother of two.

Melony Bradley is a full-time freelance craft designer/professional and a member of the Craft and Hobby Association. She lives in the small, picturesque town of Hernando, MS, where she is continually inspired to create.

Barb Chauncey is a craft designer who has contributed to several magazines and books, and is the author of *Denim by Design*. She teaches art to elementary-school students.

Jennifer Cisney is Kodak's Chief Blogger & Social Media Director by day and a crafter/gardener/knitter/photographer/cook by night. She lives in Rochester, NY, with her husband, Aaron; pug, Oscar; and cat, Stewie. Her daily adventures can be followed on her blog at www.ljcfyi.com or twitter at www.twitter.com/ljc.

Helen L. Rafson is a longtime craft designer whose designs have been published in numerous books and magazines. She specializes in kids' crafts, recycled crafts, and seasonal crafts.

Photography: PDR Productions, Inc.

Photo Stylist: Lisa Wright/Redbird Visuals

Food Stylist: Kim Hartman/Modern Amalgamated Duo, Inc.

Photo Credits:
Brand X Pictures Collection: 9, 16, 21, 22, 40, 46; **PhotoDisc Collection:** 6, 12, 15, 18, 29, 33, 34, 36, 39.

Louis Weber, CEO
Publications International, Ltd.
7373 North Cicero Avenue
Lincolnwood, Illinois 60712

ISBN-13: 978-1-4508-0635-0
ISBN-10: 1-4508-0635-X

Manufactured in China.

8 7 6 5 4 3 2 1

Contents

Crafting with Cereal

What kid doesn't love cereal? Now you can use a kid's favorite food to craft with. There are so many crafty ways to celebrate using cereal! So whether you're a mom who wants to keep her kids busy on a rainy day, a PTA mom or dad who needs a fun project for the class party, or a scout leader looking for a creative activity, you'll find something in *Clever Cereal Crafts*.

Clever Cereal Crafts offers tons of fun, with interesting projects that are designed to keep children busy the whole year through. From springtime baskets and school projects to hilarious Halloween and pretty Valentine's Day crafts, there's something for all kids' favorite holidays plus a few other delightful projects.

The simple cereal crafts in this magazine require common cooking and craft materials, many of which you probably already have on hand. Each project includes a list of instructions and materials needed. Take the time to go over the instructions for each project carefully, and be sure you have all the materials on hand before getting started. Here are just a few of the basic materials that are required for most projects in this book:

Oven or stove: When using an oven or stove, supervise children at all times.

Glue: Always use a low-temp glue gun when working with children. But even low-temperature glue guns are hot to the touch. Supervise children at all times when using any glue gun. When using craft glue, be sure it is nontoxic.

Knives: Directions may ask for something to be cut out, and a knife may be the best choice for that. If young children will be working on the project, we recommend that they use a plastic butter knife. Older children may be safe with a metal knife (not a steak or paring knife, though), but you know your children's abilities. And even with a plastic knife, it is best to always supervise children's work.

Scissors: If you are using scissors to make a project that will be eaten, be sure the scissors are washed and dried before use. Also, use appropriate scissors for the child's age. Round-tipped scissors are a better choice for younger children.

Paper: Since some of the book's projects include patterns that will need to be traced and cut out, be sure to have plenty of tracing paper on hand. When a craft project calls for scrap paper, recycle some of that computer paper and junk mail you have lying around the house!

Smock: Be sure your child wears a smock or one of your old shirts to protect clothes while working with food coloring and other messy materials.

Pattern Perfect

Some of the projects featured in this book include patterns to help you complete the craft more easily. When a project's instructions tell you to enlarge a shape, use a photocopier to do that.

Decorating with Icing

We recommend using food storage bags as pastry bags. They are inexpensive and disposable, which makes them great tools for classroom and scout projects. Another option, though, is using pastry bags, couplers, and pastry tips to decorate the crafts. These will create a more professional and polished look. If you are decorating the crafts yourself to hand out to the kids, these may be a better choice.

Some children will be able to complete the crafts with little help, but there will be times when your assistance is needed. Other projects just need a watchful eye. So it's best if you and your child review the project together and then make a decision about your role.

All the projects in this book are just ideas to get you started. Feel free to play around with the designs by changing the colors, choosing different materials, or embellishing in any number of unique ways. Let your and your children's imaginations really go wild and dream up some original projects using these as a jumping-off point. There's no limit to what they can create!

Completing the projects in this book should be an enjoyable, creative, educational, energizing experience for children. Encourage them to use their imaginations, and don't forget to praise and admire their results.

Back-to-School Bus

Get kids excited about the new school year with this fun and tasty project!

What You'll Need

- Almond Royal Icing (see page 48)
- Food coloring: yellow, red
- Cereal: shredded wheat (extra large and bite-size), chocolate cookie loops with marshmallows, fruity crispy rice, hexagon rice crisp, fruity balls
- Scissors
- Toothpick
- Fruit leather: blue, red streamer
- Student photo
- Low-temp glue gun, glue sticks (optional)
- Colorful paper plate

1 Divide icing into 3 batches; tint a batch yellow, another red, and leave last white. With yellow, ice together extra-large and bite-size shredded wheat to create bus shape. Coat entire bus with yellow icing; let harden.

2 Using icing as glue, add details. Attach chocolate cookie cereal wheels to bus, and attach marshmallows to wheel centers. Add a marshmallow headlight and orange and red crispy rice head- and taillights. Frost hexagon cereal with red icing, and use toothpick and white icing to outline and write STOP. Cut blue fruit leather into 4 windows, and add student photo to a window with icing. Attach windows to bus with icing. Add white piping around each window. From red fruit leather streamer, cut a strip and attach below windows.

3 Secure bus to center of plate with icing. (If you will be hanging plate, attach bus to plate with glue.)

4 Attach fruity balls around inside rim of plate, and attach crispy rice in a ring inside balls.

STOP
STOP

Cinnamon Trail Mix

Have kids help you make this tasty snack for their school lunches.

What You'll Need

- 2 cups corn cereal squares
- 2 cups whole wheat cereal squares or whole wheat cereal squares with mini graham crackers
- 1½ cups oyster crackers
- ½ cup broken sesame snack sticks
- 2 tablespoons butter, melted
- 1 teaspoon ground cinnamon
- ¼ teaspoon ground nutmeg
- ½ cup fruit-flavored candy pieces

1. Preheat oven to 350°F. Spray 13×9-inch baking pan with nonstick cooking spray.

2. Place cereals, oyster crackers, and sesame sticks in prepared pan; mix lightly.

3. Combine butter, cinnamon, and nutmeg in small bowl; mix well. Drizzle evenly over cereal mixture; toss to coat.

4. Bake 12 to 14 minutes or until golden brown, stirring gently after 6 minutes. Cool completely. Stir in candies. *Makes 8 servings*

Cereal Monster Balls

Celebrate Halloween with these fun characters!

What You'll Need

Cereal: fruity balls, fruity loops, oat loops
Low-temp glue gun, glue sticks
Foam balls: 3 inch, two 2½ inch
Scissors
Green fruit leather
Black frosting in tube
Small decorator tip
Red licorice strings (optional)
Ruler
2 black chenille stems

1 For Frankenstein: Use low-temp glue gun to attach green fruity balls to 3-inch foam ball. Work in a circular pattern, and continue until ball is covered.

2 Cut hair and scar from fruit leather; place on face. (Note: Fruit leather is sticky enough to adhere without glue.) Use glue gun to attach 2 purple fruity loops for eyes and 1 for mouth. Fill eyes with black frosting. Glue 2 purple cereal balls at sides of head for bolts.

3 For pumpkin: Glue orange fruity balls to 2½-inch foam ball. Work in a circular pattern, and continue until ball is covered.

4 Cut leaves from fruit leather with scissors; place on top of head. Cut a 1×2-inch piece of fruit leather; roll to make stem. Glue to top of leaves. Glue 2 purple fruity balls for eyes and 1 fruity loop for mouth. Add black frosting dot to middle of eyes. Optional: Cut licorice string to 5 inches, and glue it to top of pumpkin for hanger.

5 For spider: Glue purple fruity balls to 2½-inch foam ball. Work in a circular pattern, and continue until ball is covered.

6 Cut chenille stems in quarters for legs. Bend an end of each stem to hold cereal; thread 8 oat loops on each stem. Placing 4 on each side, stick unbent stem ends into sides of ball. Glue 2 orange fruity loops for eyes and 1 for mouth. Glue 2 purple fruity balls in center of eyes. If desired, make licorice hanger.

Note: This project uses inedible products—do not ingest! Adult supervision required.

Ghosts and Goblins

These ghosts and goblins are spooktacularly delicious!

What You'll Need for Ghosts

1 bag (10 ounces) regular marshmallows
$\frac{1}{4}$ cup ($\frac{1}{2}$ stick) butter
6 cups popcorn or puffed rice cereal
1 pound white candy coating or white chocolate, melted
24 mini chocolate chips

1 Combine marshmallows and butter in large saucepan. Cook and stir over medium heat until mixture is melted and smooth. Stir in popcorn; mix well. Use about 1 cup popcorn mixture to form each ghost shape. Allow ghosts to cool completely on waxed paper.

2 Spoon melted candy coating over each ghost to cover completely. Use fork to make ghostly folds and drapes. Decorate with mini chocolate chips for eyes. *Makes about 12 ghosts*

What You'll Need for Goblins

$\frac{1}{4}$ cup ($\frac{1}{2}$ stick) butter
6 ounces bittersweet chocolate
1 bag (10 ounces) regular marshmallows
4 cups puffed rice cereal
1 cup each cocktail peanuts and raisins
36 to 40 candy-coated chocolate pieces
18 to 20 pieces candy corn

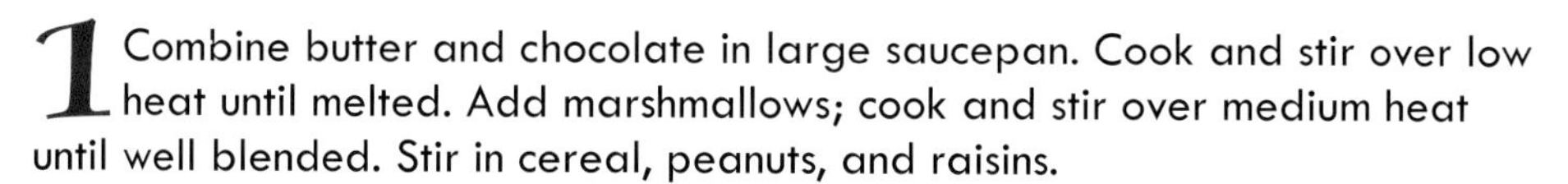

1 Combine butter and chocolate in large saucepan. Cook and stir over low heat until melted. Add marshmallows; cook and stir over medium heat until well blended. Stir in cereal, peanuts, and raisins.

2 Shape 1 cup cereal mixture into pointed figures; place on waxed paper. Use candy-coated chocolate pieces and candy corn to make faces.
Makes about 18 goblins

Gobble Gobble Granola Turkey

Your Thanksgiving dessert table will be sweeter with this fruity-loop feathered friend!

What You'll Need

Granola cereal bar
White chenille stems
Scissors
Cereal: fruity loops, fruity stars, fruity crispy rice, fruity balls
Almond Royal Icing (see page 48)
Candy sprinkles

1 Break granola bar in half, and shape half to make turkey body.

2 Cut 3 chenille stems in half for each turkey. Twist ends of 5 chenille stems to hold cereal. Thread each stem with fruity loops in following order: 3 red, 2 orange, 2 yellow. Twist together loose ends of stems to create feathers; attach feathers to back center of body with icing.

3 Attach turkey features with icing. Use yellow ball for head, tip of orange star cereal for beak, fruity crispy rice as wattle, black sprinkles for eyes, fruity loops for wings, and fruity stars for feet. Attach finished head to body.

Note: Remove chenille stems before allowing child to eat turkey!

Critter Treat Wreath

Give the birds and squirrels in your neighborhood a winter treat by hanging this edible wreath outside for them to munch on.

What You'll Need

Waxed paper
Cookie sheet
Bagel
Small spatula
Peanut butter
1½ cups oat loops cereal
Dried cranberries
12 inches ribbon

1 Spread waxed paper on cookie sheet, and place bagel on waxed paper. Use small spatula to spread peanut butter on top and sides of bagel. Gently press oat loops in peanut butter, completely covering peanut butter with cereal. Fill in empty spots with loops.

2 Turn bagel over, and cover other side with peanut butter. Cover with oat loops. Add dried cranberries throughout. Let cereal wreath rest overnight to harden.

3 Thread ribbon through bagel hole, and tie ribbon ends in a bow. Hang critter treat wreath for the birds to enjoy!

Imagination Station!

When you hang the wreath outside, look for a spot that will be sheltered from the rain and snow to make the wreath last longer.

For a Christmas wreath, use holiday oat loop cereal and red ribbon.

Festival of Loop Lights

This munchable menorah will light up your child's face with a glowing smile!

What You'll Need

Cereal: fruity loops, fruity crispy rice
Almond Royal Icing (see page 48)
Candy sprinkles: blue and yellow stars, blue and yellow balls
Mini pretzel sticks
White chocolate or almond bark, melted

1 Sort 42 blue fruity loops and 32 yellow loops. Use icing to make a column of 8 blue loops; make 4 columns. Make a column of 10 blue loops in same way. Make 4 columns of 8 yellow loops.

2 Use icing to attach columns side-by-side to create menorah shape, beginning and ending with blue and alternating colors. Place 10-loop column in center. With icing, add star sprinkles.

3 To make candles, coat 9 pretzels with melted white chocolate; while still wet, add ball sprinkles and top each with an orange crispy rice for flame. Let chocolate harden.

4 Use frosting to attach each candle to a cereal column.

Note: Adult supervision required.

Snowman Globes

Kids will love making this never-melting snowman globe—it will warm hearts of everyone dear to you!

What You'll Need (per globe)

Cereal: frosted wheat, fruity loops, miniature cookies, fruity crispy rice
Craft glue
Black frosting in tube
Small decorator tip
Fruit leather streamers
Ruler
Scissors
Construction paper: white, green or red
Paper punches: small snowflake, 1⅞-inch circle
Tall baby food jar, washed and dried
8 inches grosgrain ribbon
8 inches plastic-coated wire
Sugar

1 Stack 3 pieces wheat cereal vertically, and glue in place to make snowman. Glue a fruity loop to top of miniature cookie cereal for hat. With black frosting, make hatband where cereals are glued together. Glue hat to top of snowman. Cut ½×4-inch piece of fruit leather, and wrap it around snowman for scarf. Use frosting to make eyes and buttons on snowman. Glue on orange crispy rice for nose. Set aside.

2 Punch 8 snowflakes from white paper; glue 7 snowflakes to jar. Wrap ribbon around neck of jar, and tie it in a knot.

3 Thread fruity loops onto wire, alternating red and green loops. Insert ends of wire around ribbon tied around jar.

4 Use glue to attach snowman to bottom of jar; let glue dry. Pour sugar inside jar.

5 Punch a circle from colored construction paper, and glue it to top of lid. Glue remaining flake on top of lid. Replace jar lid.

Note: This project uses inedible products—do not ingest!

Christmas Crunch Trees

Trimming a tree has never been so yummy!

What You'll Needs

Almond Royal Icing (see page 48)
Green food coloring
Cereal: bran sticks, fruity crispy rice, fruity balls, fruity stars, woven squares
Sealable plastic bag
Sugar ice cream cones
White chocolate, melted
Licorice strings, assorted colors
Scissors

1 Tint icing green with food coloring. For needles, place bran sticks in sealable bag with green icing; coat bran sticks with icing.

2 Coat cone with icing and roll in colored needles. Use icing to attach decorations. Trim tree with assorted cereal ornaments and garlands. Top tree with cereal star.

3 For packages under tree, coat woven squares with melted white chocolate. Cut licorice strings to fit squares, and add strings for ribbon and a crispy rice bow while chocolate is still wet. Let chocolate set.

Note: Adult supervision required.

Imagination Station!
You can make an inedible, easier version of these trees. Mix white school glue with green food dye to create a green adhesive. In a sealable bag, mix bran cereal sticks with green glue. Coat cone with extra glue, then roll in green needles. Trim trees using glue and cereal decorations.

Winter Cabin Crunch

Welcome winter with this wonderland of tasty treats!

What You'll Need

½ pint milk container: emptied, cleaned, dried
Craft glue
Cereal: small frosted wheat, woven wheat squares, bran sticks, fruity loops, fruity crispy rice, fruity balls, fruity stars, chocolate cookie loops with marshmallows
Almond Royal Icing (see page 48)
Candy sprinkles
Board
Black frosting in tube
Decorator tip

1 Glue milk container shut. Coat outside of container with icing, and make roof shingles using frosted wheat cereal and woven wheat squares. Add bran sticks for siding and woven wheat squares for windows and door. Attach green fruity loop to door for wreath, and add sprinkles to wreath with icing. Use icing to make snow on housetop.

2 Create light strings with fruity crispy rice. Place cabin on board, and ice ground. Create pebble path with wheat squares and add frosted wheat snow piles.

3 Use icing to stack green fruity balls for tree. Drip icing on tree for snow. Add fruity stars to tree.

4 Using icing, stack cereal marshmallows for snowman. Top with chocolate cookie cereal and colored crispy rice. Add bran twig arms. Use black tube frosting to make eyes and buttons. Use icing to add orange star sprinkle tip for nose.

Imagination Station!
Make an entire village! Get creative, and let your mind wander over bridges, a pond, the firehouse, a school—the possibilities are endless.

SEEDS

100th Day Critters

Commemorate 100 days of delightful learning with these cute and crunchy critters!

What You'll Need

Cereal: fruity loops (100 per critter), fruity balls
Chenille stems
Large white gumdrops
Almond Royal Icing (see page 48) or nontoxic glue
Candy sprinkles
Scissors

1 For centipede: Count and sort 20 of each color of loop cereal: red, orange, yellow, green, blue.

2 Wrap end of chenille stem so loops don't fall off. Thread loops on, 10 at a time, beginning with red tail end. When chenille stem is almost completely covered, twist another stem to end to lengthen, and continue adding all 100 pieces of cereal.

3 Press gumdrop into stem end. For added strength, add icing or glue to hold. With icing, make facial details using assorted sprinkles. Cut 2 small pieces of chenille stem, and put a fruity ball on each. Attach chenille stems to head for antennae.

4 For butterfly: Count and sort 20 of each color of loops: red, orange, yellow, green. Count 10 blue and 10 purple loops.

5 For body, wrap end of chenille stem so loops don't fall off. Alternate purple and blue loops to head. Press gumdrop onto stem end. For added strength, add icing or glue to hold. With icing, make facial details using assorted sprinkles. Cut 2 small pieces of chenille stem, and put a fruity ball on each. Attach chenille stems to head for antennae.

6 For wings, wrap ends of 2 chenille stems to hold cereal. On each, thread 10 red, 10 orange, 10 yellow, and 10 green loops. When done, twist loose ends together to create teardrop-shape wings. Attach wings to body with another chenille stem. Add a spiraled chenille stem bottom wing to each side.

Note: If using glue, do not let children eat project! If using icing, remove chenille stems before eating.

Panning for Gold

This tasty project would be a fun accompaniment to a lesson on the gold rush!

What You'll Need

6 cups chocolate crispy rice cereal
3 cups chocolate cereal balls
2 cups puffed corn cereal
½ cup broken chocolate wafer cookies or chocolate sandwich cookies (about 14), cream filling removed
½ cup miniature chocolate chips
Large bowl
Small bowl for each child

1 Combine all ingredients in large bowl.

2 Have each child scoop out ½ cup mixture from bowl and empty onto his or her own plate. Allow each child to remove and count out the puffed corn cereal pieces ("gold nuggets") from the "dirt" and "sand" on his or her plate. The child with the most gold nuggets wins! *Makes 24 servings*

"O"s and Bows Heart

Let your sweethearts know how sweet you think they really are with a cereal and licorice heart card!

What You'll Need

Cardstock: red, white
2 different decorative-edged scissors
Licorice strings, 12 inch lengths
Red fruity loop cereal
Pencil
⅛-inch paper punch
2 red twist ties
Craft glue

1 Using patterns below, cut larger heart from red cardstock with decorative-edged scissors and smaller heart from white cardstock with other decorative-edged scissors.

2 Lay 2 licorice strings side by side, and tie them together with a knot close to one end. String fruity loops onto each side of licorice, leaving enough licorice free at ends to tie a bow. (Small children will need help tying bow.)

3 Lay cereal heart on white paper heart. Use pencil to lightly mark placement of holes at sides, top, and bottom of heart. Punch holes in white heart at marks. Cut twist ties in half; use twist ties to attach licorice heart to white heart, twisting ends on back of white heart. Glue white heart to red heart.

Enlarge patterns 200 percent.

You're so
sweet,
VALENTINE!

Lucky Leprechaun Hat

Lucky are the lads and lassies who try these tasty toppers!

What You'll Need

Green wafer ice cream cone
Almond Royal Icing (see page 48)
Plastic knife
Cereal: fruity crispy rice, oat bran squares, bran sticks
Rainbow-stripe fruit leather streamers
Scissors

1 Place cone upside-down on work surface. Coat bottom brim of cone with icing using knife, and place green crispy rice cereal around brim.

2 Use icing to attach rainbow fruit leather above brim; trim excess fruit leather. Add oat square and bran stick buckle with more icing.

3 Add green crispy rice to top of cone with icing.

Imagination Station!

Make an inedible version using glue instead of icing. (Be sure to use nontoxic glue, and be sure children don't eat finished craft!)

Ice a cupcake to look like a leprechaun's face, and place hat sideways on the plate to top his head for an incredibly cute party treat!

Crispy Kaleidoscope Eggs

Make springtime even more fun with these crispy and colorful eggs!

What You'll Need

3 tablespoons butter
3 cups mini marshmallows
2½ cups crispy rice cereal
2 cups fruity loop cereal
½ cup jelly beans

1 Microwave butter in large microwavable bowl on HIGH 1 minute or until completely melted.

2 Add marshmallows, cereals, and jelly beans to bowl; microwave on HIGH 1 minute. Stir gently to combine ingredients without crushing cereal. Let stand 2 minutes or until cool enough to handle.

3 Butter hands well, and form ¼ cupfuls of cereal mixture into egg shapes (about the size of real eggs). Mixture will remain moldable for about 10 minutes.

Makes about 2 dozen eggs

Note: Adult supervision required.

Silly Cereal Self-Portraits

Display these silly cereal self-portraits, and watch the smiles appear!

What You'll Need

Pencil
Cardstock: assorted tan and beige shades
Scissors
Craft glue
Assorted scrapbook paper sheets
Cereal: fruity loops, fruity crispy rice, plain crispy rice, bran sticks, fruity balls, fruity stars, chocolate cookie loops, small squares
White candy sprinkles

1 Use pencil to draw a face/neck shape and 2 arms on desired color cardstock. Cut out shapes, and glue them on selected scrapbook sheet.

2 Glue cereal pieces to face to add details. For example, for eyes use loops, add fruity crispy rice to center hole, and add white sprinkles for reflection. But let children use their imaginations to create their faces.

3 Glue on hair using desired cereal.

4 Lightly draw shirt shape and other design patterns with pencil. Glue cereal to create apparel and accessories.

Note: This project uses inedible products—do not ingest!

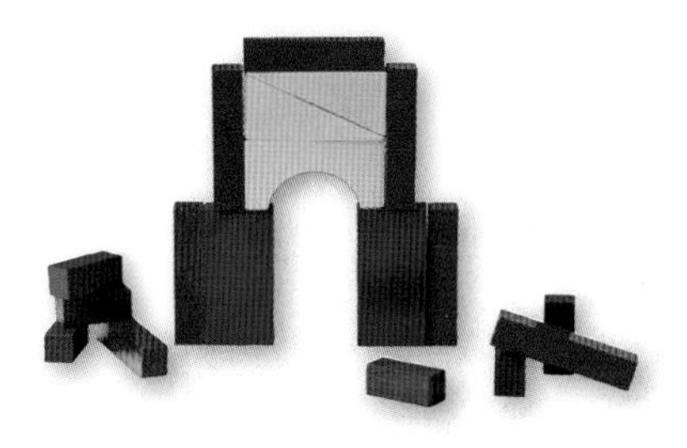

Disney
Friendly Stories
Pony
Spider's SING-ALONG STORY
MY FIRST 1000 WORDS
Farm Animals

Springy Cereal Baskets

These fun baskets are quick craft projects for kids of all ages.

What You'll Need

8-ounce foam cups
Fruity loop cereal
Craft glue
White chenille stem
Ruler

1 Using photograph as a guide, glue fruity loops onto foam cups in pattern desired. Let kids to use their imaginations to create their own designs, if they so choose. Let glue dry.

2 String loop cereal onto chenille stem handle, leaving 1½ inches on each end without cereal. Glue first and last cereal loops to chenille stem ends to hold cereal in place. Let dry. Glue ends of handle to inside of cup. Let dry.

Note: This project uses inedible products—do not ingest!

Mommy's Special Day

Show Mom how much your love blooms with this sweet showstopper!

What You'll Need

- Cereal: honeycomb shapes, fruity stars, fruity crispy rice, fruity loops, fruity balls
- Nontoxic acrylic paint in desired colors
- Paintbrushes
- Colored cardstock
- Scissors
- Double-sided tape or nontoxic glue

1 Paint honeycomb cereal pieces in desired colors; let dry.

2 Cut 3 pieces of assorted cardstock in graduating sizes, using double-sided tape or glue to attach pieces, creating frame.

3 Create flowers with honeycomb-shape centers and petals, stems, and leaves of assorted cereals, and glue to cardstock.

4 Use fruity crispy rice and fruity loops to spell "MOM." Glue colored cereal balls around outside of frame; let dry.

Note: This project uses inedible products—do not ingest!

You're Invited
call
MOM
notes
Oatmeal Raisin Cookies
(2 sticks) margarine or butter
firmly packed brown sugar
teaspoon vanilla
cups flour
teaspoon baking soda
teaspoon salt (optional)
teaspoon cinnamon
3 cups quick oats
1 cup raisins
1 cup chopped nuts (optional)
file

Dad, you're toadily cool!
Dad, I love you beary much!
COOPERSTOWN
THE LOVE OF
ASEBALL
The Wit & Wisdom of Baseball

Dad's Day Greeting Cards

Make Dad a card to express your sweet Father's Day wishes!

What You'll Need

Construction paper: brown, blue, green, purple
Ruler
Pencil
Scissors
2 white cards, 5×6½ inches each
Paper glue
Black permanent marker or computer
Chocolate graham crackers
Frosting in tube: black, white
Small decorator tip
Cereal: chocolate chip cookies, fruity loops
Sealable plastic bag
Small hammer
Small bowl
Paintbrush

1 For bear card: Cut brown construction paper to 5×6½ inches. Adhere to front of white card with paper glue. Cut blue paper to 3½×4 inches. Adhere to top center front of brown card.

2 Write "Dad, I love you beary much!" at bottom of card front. (Note: If using computer, print message on brown paper before cutting and gluing to white card.)

3 Ice 1 side of graham cracker with frosting, and place on front of card on blue square. Use miniature cookie cereal and black frosting to decorate bear and line sides of blue paper.

4 For frog card: Follow step 1, using green and purple construction paper. Follow step 2 to write "Dad, you're toadily cool!" at bottom of card front.

5 Crush about 20 green fruity loops in plastic bag with hammer. Place 1 tablespoon white frosting in small bowl; dilute frosting with 1 teaspoon water. Brush frosting oval on graham cracker square, and sprinkle with crushed green cereal. Attach fruity loops with black frosting for eyes. Adhere graham cracker with frosting on front of purple square.

6 Use icing to attach alternating colors of fruity loops around graham cracker.

Cereal Jungle

Let kids' imaginations run on the wild side with these fun animals!

What You'll Need

Cereal: fruity loops, oat loops
Bumpy chenille stems: 2 green, scrap red, 3 brown
Scissors
Ruler
Wiggle eyes
Craft glue

1 For snake: Thread fruity loops on green bumpy chenille stem in repeating pattern (for example: green, red, yellow, orange, purple; repeat pattern). Make loop on an end of stem to secure cereal.

2 Cut remaining green stem to 5 inches, and thread 6 fruity loops onto stem. Form a ring, and twist ends together to secure. Cut 1 bump off green chenille stem piece. Twist ends on each side of ring to secure, making sure 4 cereal loops are on top and 2 are on bottom of ring. Glue 2 wiggle eyes onto center bump. Twist end of red scrap stem to bottom of ring for tongue. Bend end. Attach snake body to head.

3 For monkey: Cut 1 brown stem in half, and thread 10 oat loops onto half; place 5 loops on each half of stem. Bend each end to form hands of monkey. Set aside.

4 Thread 6 oat loops on other half of stem. Form a ring with stem, and twist ends to secure. Place all cereal pieces on half of ring. Cut 2 inches from second stem. Thread an oat loop on each end of piece for ears. Place piece on middle of ring, and twist ends on each side of ring to secure. (Keep all loops on top of ring.) Glue 2 wiggle eyes onto center piece. Glue an oat loop on bottom of ring for mouth.

5 Cut 4 inches from second brown stem, and thread 5 oat loops on stem. Twist end of stem on monkey head to form torso. Attach arms at top of torso. Cut 3 bumps from stem, and twist end around torso to form tail. Curl tail end. Attach middle of last stem to bottom of torso. Thread 5 oat loops on each end, and bend ends to form feet.

Note: Remove chenille stems before allowing child to eat cereal!

Mini Masterpieces

Let your little Picasso make fun masterpieces by painting with cereal.

What You'll Need

Fruity loops cereal
Small sealable bags
Small hammer
White Decorator's Icing
(see page 48)
Small bowl
Chocolate graham crackers
Paintbrushes (dedicated to kitchen crafts only)

1 Sort colors of cereal into small plastic bags. Use hammer to crush cereal into fine powder.

2 Place 1 tablespoon icing into small bowl. Add 1 teaspoon water, and mix until blended.

3 Break graham crackers into halves. Use paintbrush to paint graham crackers with icing, then sprinkle area with cereal crumbs. Work with 1 color at a time to prevent areas from bleeding together.

Almond Royal Icing

2 egg whites*
4 to $4\frac{1}{2}$ cups sifted powdered sugar, divided
$\frac{3}{4}$ teaspoon almond extract

**Use only grade A, clean, uncracked eggs*

1 Beat egg whites in medium bowl with electric mixer at high speed until foamy.

2 Gradually add 4 cups powdered sugar and almond extract. Beat at low speed until moistened. Increase mixer speed to high and beat until icing is stiff, adding additional powdered sugar if needed.

White Decorator's Icing

2 egg whites*
$\frac{1}{2}$ teaspoon cream of tartar
4 cups sifted powdered sugar

**Use only grade A, clean, uncracked eggs*

Combine egg whites and cream of tartar in large bowl. Beat with electric mixer at medium speed until foamy. Beat in sugar, $\frac{1}{2}$ cup at a time, beating until very stiff. Use within 1 hour. Cover with damp, not wet, paper towel until ready to use. (Icing is best for piping decorations you want to completely harden.) *Makes about 2 cups*